"

You can dream, create, design, and build the most wonderful place in the world. But it requires people to make the dream a reality.

—WALT DISNEY

"

Each of us is carving a stone, erecting a column, or cutting a piece of stained glass in the construction of something much bigger than ourselves.

—ADRIENNE CLARKSON

"

Use the best possible materials, and reveal the quality of those materials and the craftsmanship of their assembly.

—KARL FRIEDRICH SCHINKEL

A BUILDER'S LIFE DONE WELL

THE STORY OF PRUTTING + CO.

A REFLECTION ON FIFTY YEARS OF BUILDING

BY DAVID L. PRUTTING

images
Publishing

CONTENTS

FOR DEBORAH

September 2025 would have marked 50 years married for Deb and I; she passed away in August 2023. She chose to support my career wholeheartedly rather than pursue her own professional interests, in which she would have succeeded immensely as she was very talented aesthetically with design, crafts, and the arts. She started by holding the ladder for me, then overcoming her fear of heights to climb the ladder with me, and then diligently hammering down roof shingles. Eventually she became a devoted mother to our two children and a strong community contributor, but she anchored our office efficiently for many years as our business grew. As Deb was a very principled person, I relied on her to be my moral compass and followed her instincts about calling a spade a spade. No nonsense from this lady and you knew where you stood.

Certainly, Prutting + Co. would not be what it is today without her guidance and efforts. So many, many thanks and thinking of you fondly Deborah as I dedicate this book to memories of you and your dynamo spirit!

ELIZABETH WITH DEBBIE, 1995

DEBBIE, 1980

DEBBIE, 1977

INTRODUCTION

I am struggling to get started here as I'm searching for the rationale to write this book. It's about a career filled with projects, incidents, and experiences, and the story is told well by the photos, photos, and more photos. I look at these, and the scope and challenges of so many of these jobs sometimes floors me. We did that? How? What type of people commission these projects, and where do they find the talented designers/architects to conceive them, and the courage to trust and turn them loose on the designs? I am fortunate to have been in the middle of so many of these jobs. To use a filmmaking analogy, the clients are the producers: they hire and invest confidence in the directors (architects), and we—the builders—provide the players to make it happen. Of course, there are plot twists, drama, and often lots of emotion. I do not encourage most people to take these challenges on, but fortunately, they ignore that advice, and we've had the opportunity to create some amazing buildings.

It is from that basis that I will share some advice, wisdom, and warnings, and tell the interesting stories of how things happened. I have several specific target audiences in mind here—brave people who want to build, architects and designers, builders like myself, and others who may not be so interested in those details but will enjoy looking at some stunning pictures.

A little personal background first. I'm from Ellington in northeastern Connecticut, and one of four sons. Dad was a musician, and most of his career revolved around music. Mom was a teacher who taught multiple grades in a one-room schoolhouse. Of his four sons, my dad would later tell me I was the one least likely to become a carpenter, as I was clumsy, dropped a lot of things, and was useless with a screwdriver.

I entered Syracuse University around 1968, spent three years there, played some football, and left during a tumultuous time of protest and cultural upheaval. An English major, I made a lot of friends, learnt a bit about a variety of things and people, and experienced an existential broadening. Forgoing a degree, I went on the road for a year or so before settling in Cape Cod, MA, where I was hired by a local homebuilder named Arthur K. Finlay. His family had reportedly been anchored in Rock Harbor, Orleans, for 17 generations. My primary attribute at that point was being physically capable of moving a lot of lumber around—someone colloquially known as a lumper—and that is how I broke in.

It was a good group of four or five guys, and we did everything on these modest Cape-style houses: framing, roofing, sidewall, including lots of interior details like building and finishing staircases, trim millwork, and gable louvers. It was only years later that I learnt that premade vanities, kitchen cabinets, and staircases are available to buy. It was a valuable apprenticeship, and I became proficient at sidewall shingling and roofing, so that four years later my first business card said "Custom wood roofs."

Business slowed around 1975 and, wanting to save some money so I could get married, I left the Cape bound for Houston, TX, which I was told was jumping with construction jobs. Fortunately, I had a cousin who lived in Houston, and she housed me initially. Carpenter jobs were plentiful, so I jobbed around, worked as a waiter in the evenings, and even performed a little with my guitar. Around three months later, I married a dynamic, young, and remarkably talented woman named Deborah Ann DiMicco. We enjoyed a brief honeymoon, and then it was back to Houston for work, where I had made good connections and work was still readily available. Deb, from a bucolic part of Westchester County, NY, cried when we drove through the flat, barren landscape of Beaumont, TX, on our way to Houston.

I began a wood-shingle roof job in an aging upscale subdivision, and as neighbor after neighbor called for my services, Domino Roofing was born. Deb, not wanting any part of pursuing career interests in Houston, had latched on as a roofer. Able-bodied and nimble, she was quite the producer while I did most of the heavier lifting. We became quite an attraction—a husband-and-wife roofing team—and that notoriety got us a lot of work. I enjoyed working for the straight shooters and honorable people there; I never experienced any bad outcomes. One potential client posed this question to me: "Down here, only two types of people do this type of work—snowbirds and jailbirds. Which one are you?"

My growing client list for roofing led to other work—I was asked to build decks and garages, and do miscellaneous carpentry work. We opened an art gallery in a spacious 2000-square-foot showroom, where we also lived. We had to push the rollout bed back in place before heading to work, then sneak back in after a sweaty day's work on the roof. With Deb managing the gallery, I at least got her off the roof. We were fortunate to know many talented outer Cape artists who trusted us to represent their work, and this connected us with a nice art scene in Houston and around the state. The artworks were better than our represention of them, but overall, it was a fun though not profitable endeavor.

Returning east around 1980, we eventually settled in New Canaan, CT, near Deb's family in Westchester. I jobbed around, advertising carpentry work in the local newspapers. I landed a bigger project working for two architects after they fired the contractor who had hired me to work there with him. As it was a complete house renovation so I was there for a while, and the architects then hired me to renovate the nearby Vista Market, a grubby beer and raffle-ticket buying place. I was a one-man team. I recall pushing shopping carts filled with rubble down the grocery aisles to a dumpster outside, and one memorable time cutting through the flat roof to accommodate a stairwell, thinking, "Here we go! No turning back, gotta finish this up quick!" By now my little business was expanding, and I was hiring subcontractors and learning how to manage them while also dealing with the clients—I had to be a carpenter and a diplomat. There's little worse than invading someone's home, bringing a fleet of vans, and dominating the driveway, and the stress that can cause. Perhaps the biggest stressor is asking for, and sometimes demanding, the decisions needed to keep construction progressing.

THE SCOPE AND CHALLENGES OF SO MANY OF THESE JOBS SOMETIMES FLOORS ME. WE DID THAT?

Having worked many different jobs, from mowing lawns to steel-mill assembly line to pumping gas, I did relish self-employment. I ran a remodeling company for 15 years: small jobs, whole house remodels, whatever I could find. I assembled a good group of subcontractors and

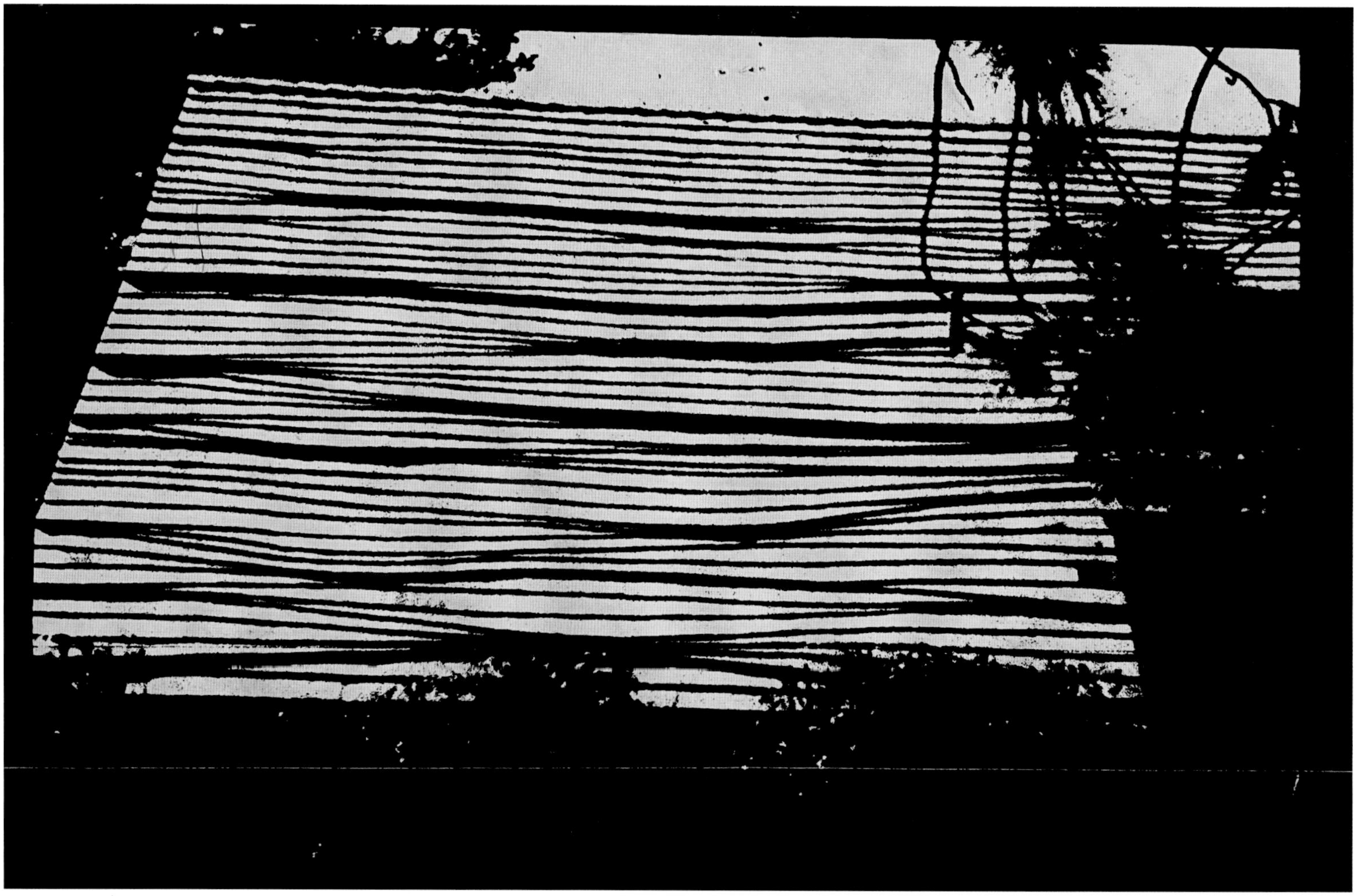

learnt a lot from them. Eventually we built a very nice new custom home, moving us into a bigger market. In many ways, these ground-up jobs were far simpler than attacking the clutter and mess of remodeling someone's existing home and dominating their driveway. Seeking new business, I found it hard to compete with many local, established McMansion builders who had extensive portfolios. I sought a market niche and found it after purchasing a run-down "modern" house in New Canaan. This was our third purchase in town as we utilized the capital gains exemption by fixing up our two prior residences, then selling them for profit.

PRUTTING CREW, 1987

This decrepit modern house sat on a gorgeous 3-acre spread with a pool and a newly built pool house that was quite habitable, so we put our two teenage kids there while Deb and I roughed it in the main house. Living apart from the kids, we often used phone lines to call them up for meals. Plaster walls crumbling, seals shot on windows, radiant copper piping failing—the seller even took the sconces and soap dishes and left us with dangling wires and holes in the walls. We interviewed several architects as we contemplated a renovation or a teardown. This was during an era of heightened local consciousness about "How do we save the Moderns?" A significant part of New Canaan's architectural heritage originated in the 1950s and 1960s, as the Harvard Five architects, led by Marcel Breuer and Phillip Johnson, all lived and worked there, designing many simplistic, utilitarian, low-cost, unadorned, modest-sized houses—smartly sited on what would later become very expensive lots.

We hired Joeb Moore, whose accomplished architectural firm I had worked with before on several projects. As I knew he would, Joeb and his talented team brought great energy and enthusiasm to the project. Our "modern" ended up being a teardown and build back on the same footprint, with a new second story "floating" above the original ground-floor building. It maintained an homage to the original design but was startlingly new, bigger, and modern—a unique and clever response to the modernist teardown dilemma. The house was published and much discussed, and soon I had a new group of architects calling, as we had acquired a reputation for understanding the detailing for trimless minimalist houses.

With a broadening client base, expanding portfolio, and a well-earned reputation, we welcomed the business of all kinds that came to our door. I liken growth for a small business like ours to an upside-down triangle—contacts and client base constantly widen over time. With hard work, dutiful customer service, and great moral and ethical guidance from my partner and deeply principled wife, Deb, the company has gone on to accomplish a great deal and been involved in many spectacular projects with great people.

FOR THE CLIENTS

FOR THE CLIENTS

Many of my clients produce, and determinedly build, houses with the same enthusiasm as collectors of fine art or cars. But they must be courageous and, of course, wealthy enough . . . and patient. Every project seems to take longer and cost more than initially expected. If you're looking for land to build new, good lots are difficult to find in established, developed areas, so the best option may be to find existing homes nicely sited and located to either knock down or modify. The landscape, trees, and utilities can be reused and integrated into new plans to reduce costs.

I'm a dedicated proponent of good design, which, of course, comes in many shapes and sizes. Quality, thoughtful design will hold value and, importantly, provide comfort and aesthetic satisfaction. These are the most invaluable investments you will make. The choosing of an architect is like a marriage—trust and lifestyle intimacies are shared. It is a partnership, so you need to be comfortable with your choice. It's true that some architects can persuade you to create their vision rather than yours, which might be the better way to go. However, make sure you get what you need while being open to change. I believe that it's easy to second-guess an idea, but not so easy to originate it. I've seen many millions of dollars spent on mediocrity, so when interviewing, I recommend being diligent and open to ideas. Then buckle up and keep the faith because these projects are often marathons, not five-mile races.

When it comes to choosing your design team, I'd suggest multiple interviews and reference checks—the same for hiring builders. Remove any builder from consideration who advertises as always delivering "on time and on budget." These complex projects often start too soon with no realistic time frame, and plans change all the time. Generally, the architect initially is the hero, then later becomes the bum, then glories in the finished product. Meanwhile the builder carries the day through the rough spots, and then when the client spends too much and patience wears out, becomes the bum.

THE GREATER THE PURCHASING POWER OF THE OWNER, THE TOUGHER THE CHOICES AND DECISIONS CAN BE.

It can be a strain, but it typically ends well, as ideally it should, because the client needs us to guide them in how to use and maintain these new complex buildings.

The greater the purchasing power of the owner, the tougher the choices and decisions can be. The ability to afford almost anything makes options unlimited. Making prompt, timely decisions is the number one requirement for keeping jobs running smoothly and close to optimal schedules. It's a big responsibility and can be overwhelming. This is where faith in the design team comes in. It's their studied profession, so trust them. Then get ready to hire and spend—realtors, lawyers, engineers, consultants, surveyors, contractors, permits, and more.

Sometimes the builder and subcontractors pop their heads in here, as practical feedback is also essential. A smart architect is open to this input and often seeks it. We had a general manager named Dave McGeehan with a wealth of building experience, whose recommendations I told clients were the right choice 19 out of 20 times and should always be seriously considered.

But the owner/client is the producer and the ultimate decision-maker. Make sure your team knows what you're feeling and thinking. Perhaps my favorite client was a man named Burton Resnick, a NYC real estate developer who was, shall we say, a great motivator. Meetings were direct, decisive, and ended with rejected crumpled written proposals all over the floor. Phone calls rarely lasted more than ten to twelve seconds, and directives were made emphatically clear. I had to be prepared for those moments, as no time-wasting was allowed or tolerated. I learnt a lot from this man.

This advice applies to all levels of new construction and remodeling, which can be a brutal experience. Invasion of your occupied living space requires lots of patience, and the team you hire better be motivated to get it done. Your goal should be to seek practical solutions that look good and work well.

HORIZON HOUSE

This was a difficult project on a hillside with a 45 percent slope directly above the train tracks next to the Hudson River. The train line required a $250,000 bond and posted an observer on-site prepared to alert the trains in the event some equipment rolled downhill and onto the tracks. The plan called for a central building with several bedrooms, and an adjacent attached elevated concrete viewing platform supported by two plinth slabs. This required a forest of steel rebar and some great engineering. Seventy-two pilings were necessary as we cut into and held back the hill and used the out-fill to make a platform for heavy equipment to operate. This project, like several others, was beyond my personal expertise, but the costing and building were performed successfully by our smart and courageous staff. It's a remarkable design, and the talented Holl team was thorough and kept everyone on task in a pleasant and professional way.

ARCHITECTURE: STEVEN HOLL ARCHITECTS

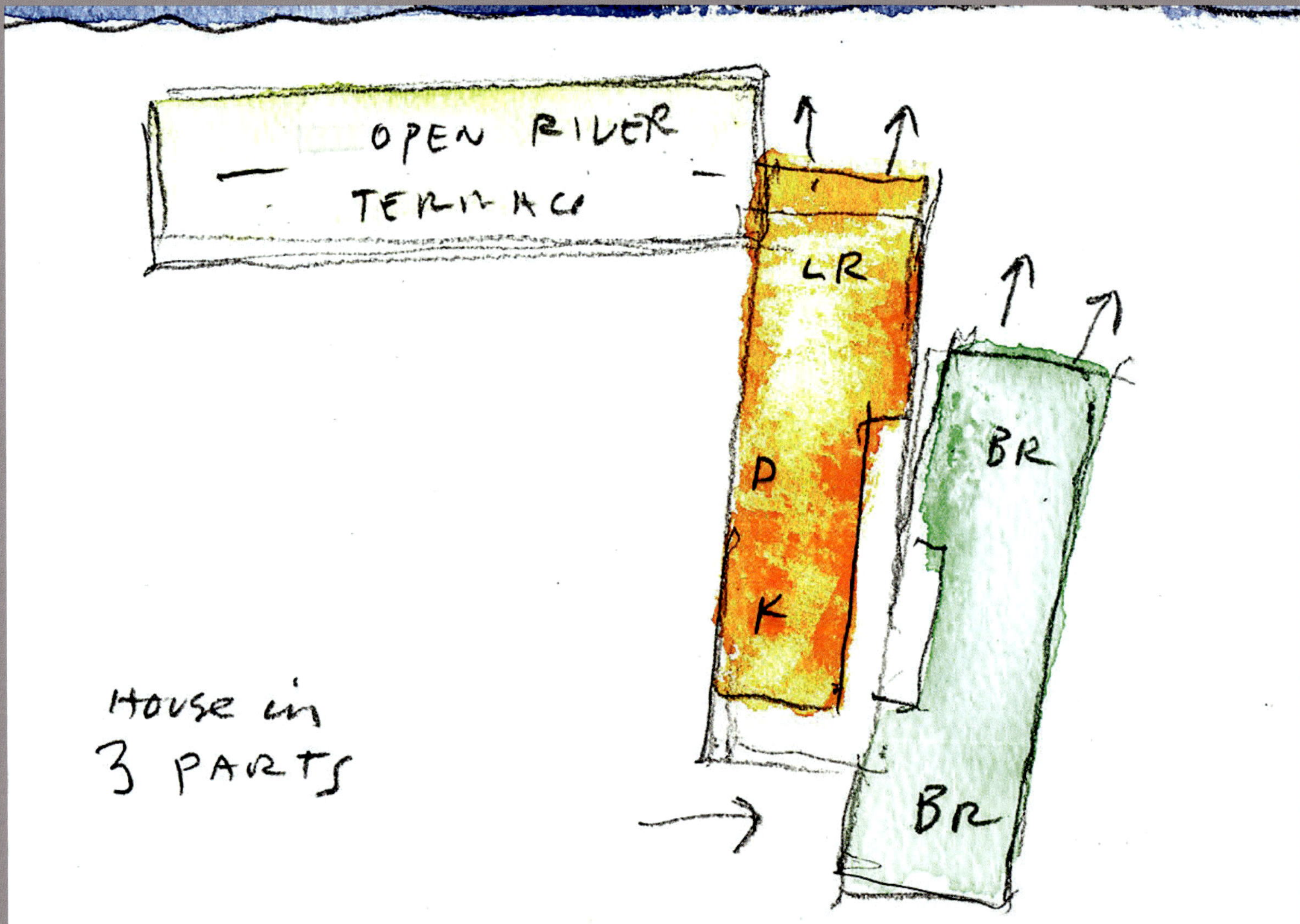
OPEN RIVER
TERRACE
LR
D
K
BR
BR
House in
3 PARTS

GREENWICH HARBOR HOUSE

A very sleek contemporary house squeezed onto an undersized lot, yet made to feel surprisingly private. Featuring an indoor atrium and perfectly proportioned rooms, the large wraparound porches provide great waterfront views. Terraced down a hillside, the thoughtful landscaping provides a Zen-like ambience using quiet, serene entry-area plantings.

ARCHITECTURE: AMANDA MARTOCCHIO ARCHITECTURE

A HOUSE IN GHENT

Entered through a newly cleared mile-long driveway, this challenging project comprised four separate glass boxes on a stony crest connected by elevated heated concrete walkways. Utilities ran to each building underground, centralized in one basement utility space. The structural engineering was stringent, and the buildings were skinned with high-performance curtain-wall glazing. The interiors were sparsely decorated and flooded with daylight, and had spectacular views of the countryside. The owners were supportive, friendly, and clearly visionaries. The finished product is quite amazing.

ARCHITECTURE: TOSHIKO MORI ARCHITECT

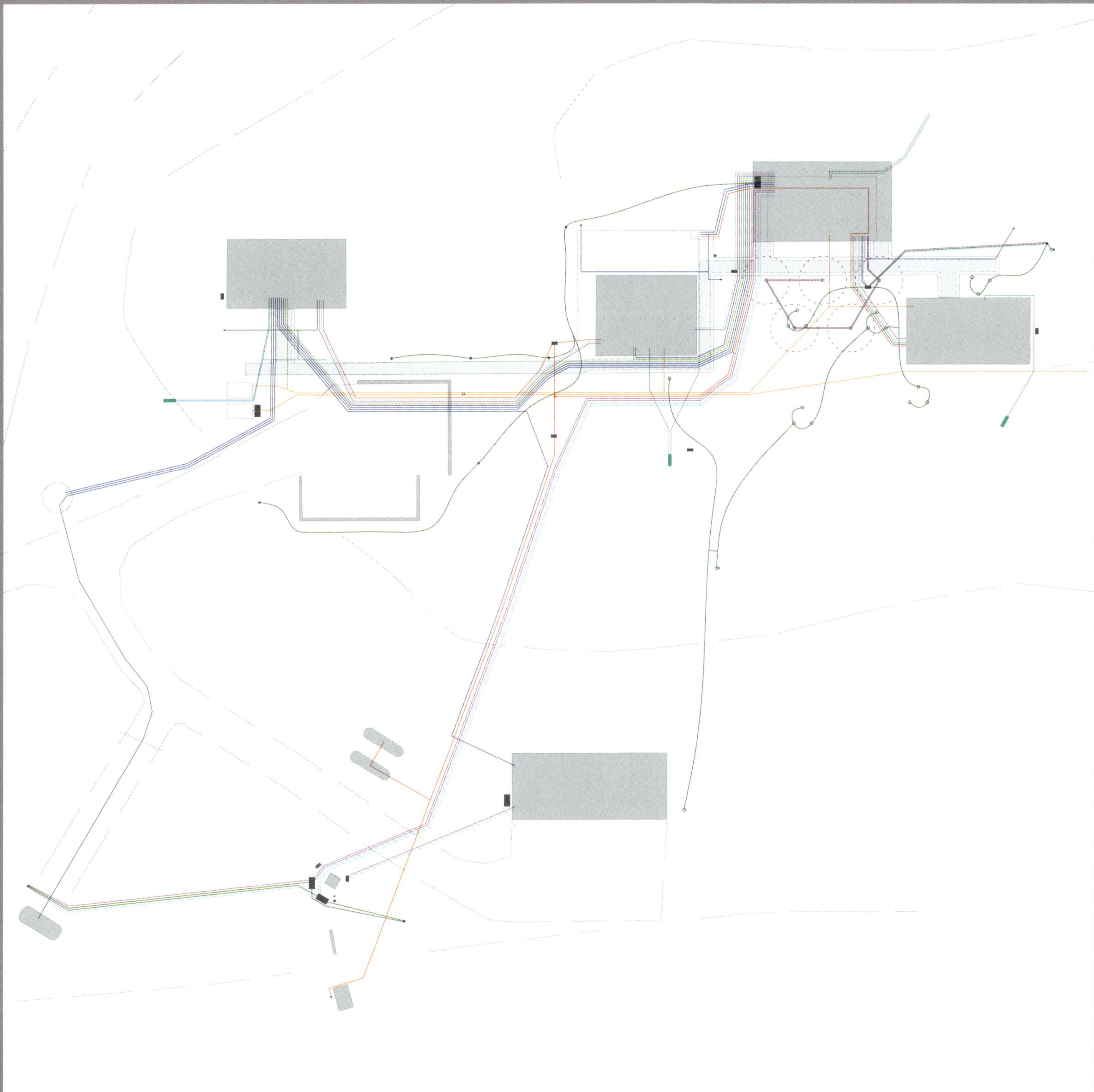

DRAWING BY PRUTTING + CO.

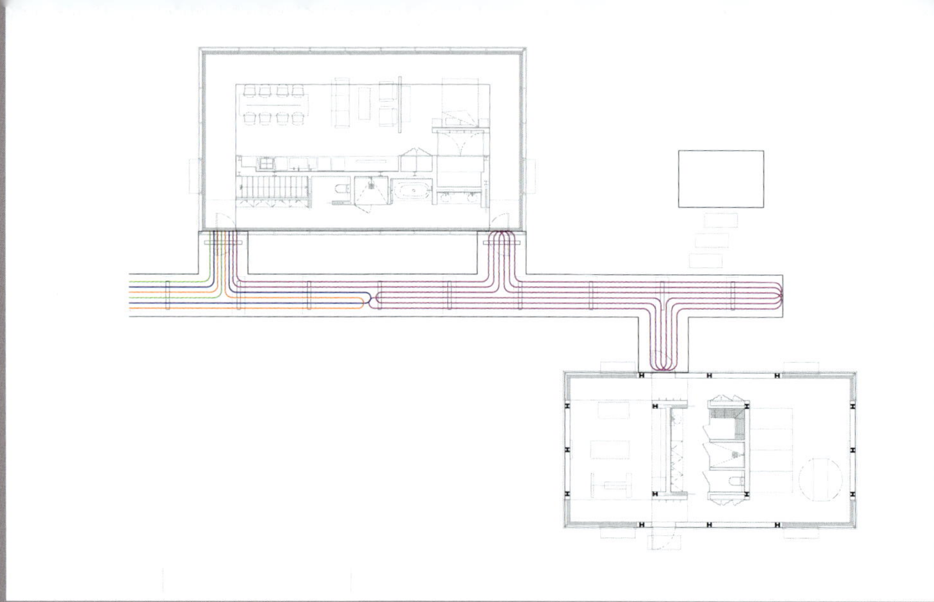

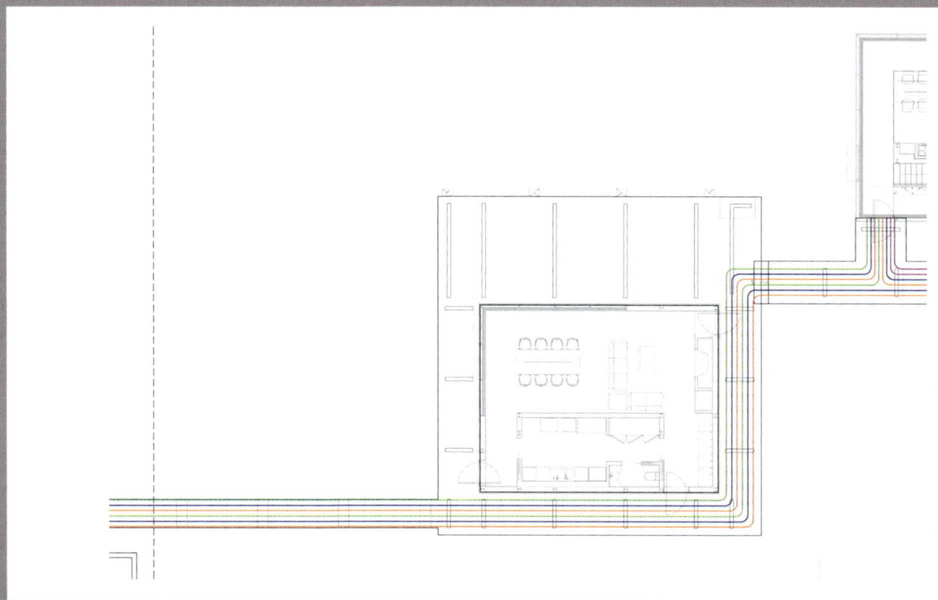

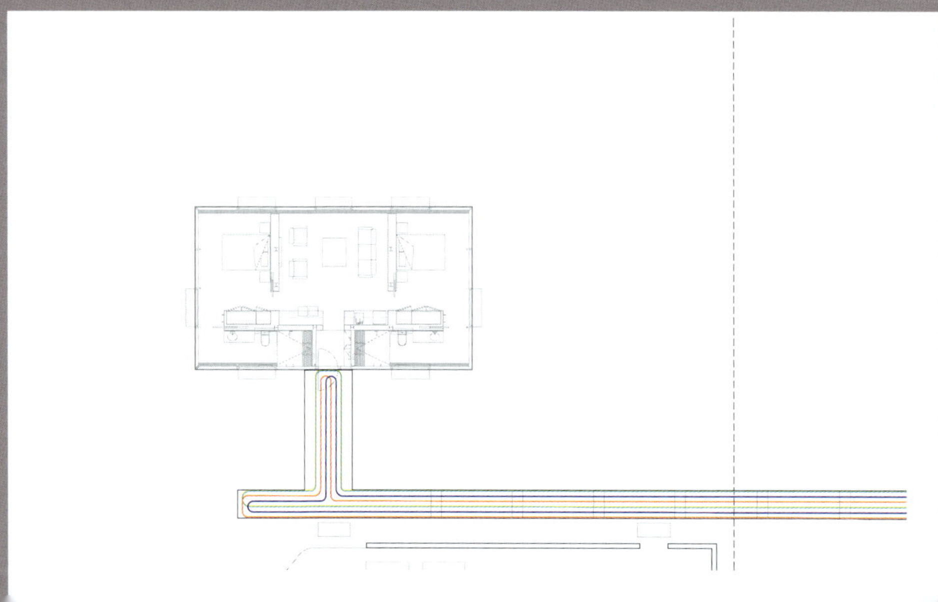

HEATED WALKWAY PIPING
DRAWINGS BY PRUTTING + CO.

RIVER HOUSE

This was an addition/renovation to a contemporary house in Connecticut. Included was a new, largely glass wing and renovations throughout. An elaborate staging was erected to do extensive work on the existing building portion over the river. The interiors turned out stunningly cool: sleek, minimal, yet loaded with style and a perfect backdrop for the owner's contemporary art collection.

ARCHITECTURE: JOEB MOORE & PARTNERS

HOCKNEY

NEW CANAAN TOWNHOUSE

Talk about courageous—in this case, it was me and, somewhat reluctantly, my wife, building our only spec house in 2008. I wanted to build a house of the future, incorporating as many hip, smart features and systems as we could afford, such as photovoltaics, solar thermal, LEED certification, and more. The design shape smartly evolved from and conformed to zoning regulations: setbacks, height limits, etc. I wanted a swimming pool on the roof, so the entire structural skeleton was built in steel. Proud of it, I put a huge price tag on it and, of course, nobody bought it, so we ended up moving in for ten-plus years. Trying to generate interest, I had spread a rumor that Derek Jeter was looking at it as the perfect bachelor pad. He was not.

ARCHITECTURE: JOEB MOORE & PARTNERS

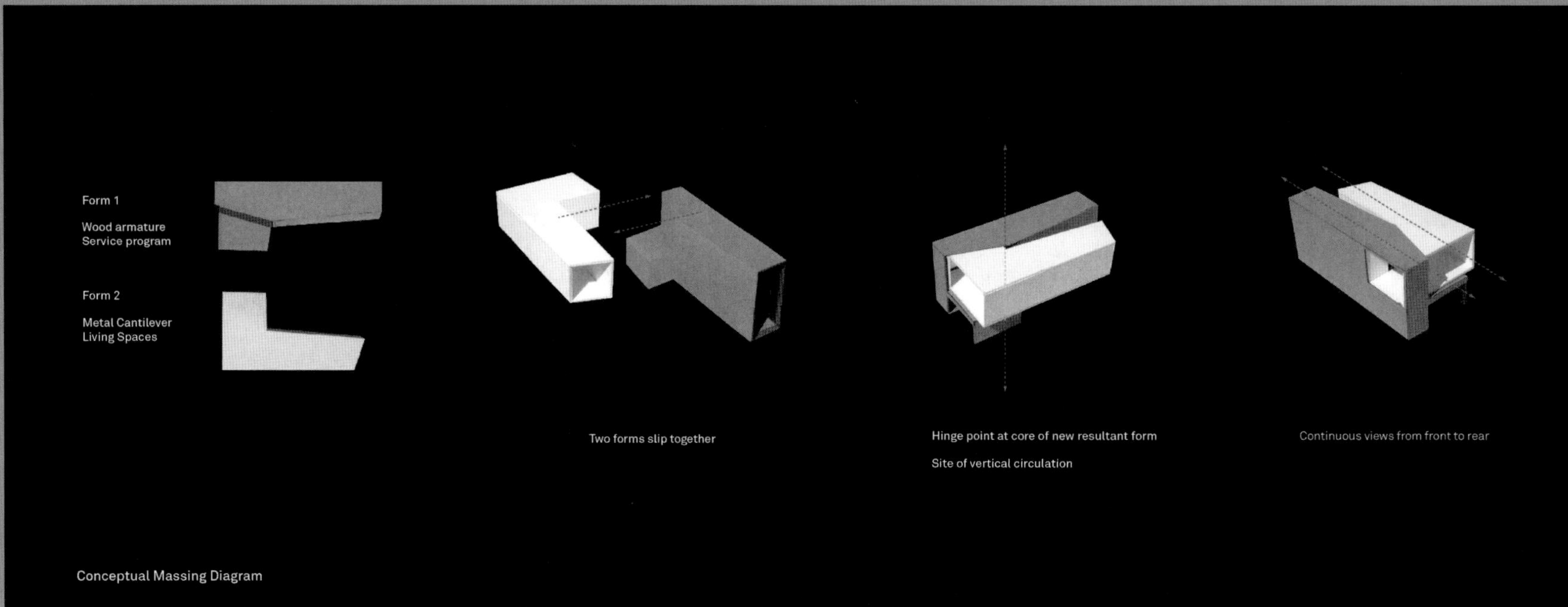

Conceptual Massing Diagram

MODELS AND PLANS BY JOEB MOORE & PARTNERS

FOR THE ARCHITECTS

FOR THE ARCHITECTS

Throughout my career, I've worked with many different architects from all over the country. We builders bridge the gap between drawings and reality. Our intention is always to make the architect look good—not only by good construction performance, but sometimes by uncovering mistakes and devising solutions. Early in my career, I dutifully ordered window units off a plan where one critical unit was incorrectly specified. The desired rough opening was built, and the window had to be re-ordered. I remember the architect visiting at the owner's request and later whispering quietly to me: "Just fix it, Dave."

Many of these experiences were memorable, and almost all were good. I built two houses with Jim Cutler of Cutler Anderson Architects from Seattle, who hand-drew all the critical details of a house—interior and exterior, desired intersections, etc. These were all put together in a specification book, divided logically into sections. He told me he drew them on plane rides while traveling to various projects. Jim is a bit eccentric and asked me not to speak to clients even when they asked questions; instead, I was to call him about anything. I would describe him as one of the most scrupulous professionals I've dealt with, as he once reminded me while discussing a financial situation in which he represented the owner, and though objective, he had the client's best interests in mind. Jim also sometimes camped in the finished houses for several days, "test-driving" them to catch any remaining issues and develop a punch list before turning it over to the clients.

I recall a memorable incident after pricing a very challenging project with Steven Holl Architects. I was summoned to a meeting where Steven himself, with a black marker in hand, lopped off about one-third of a floor plan to save millions of dollars, and made me sign the amended plan with a reduced budget spelled out. Simple as that, and it got built.

Unfortunately, I've witnessed insensitive clients questioning architects' invoices in front of me while leveraging them to discredit my extra bills. As my friend and talented architect Mark Demmerle would tell me, he regularly had to "dance for his supper." I had some experience with this on the West Coast in Portland, where I partnered with a great former employee named Andrew Friedman on some building projects. It is a great training ground for architects, as there are no lavish budgets, and designers usually have to make

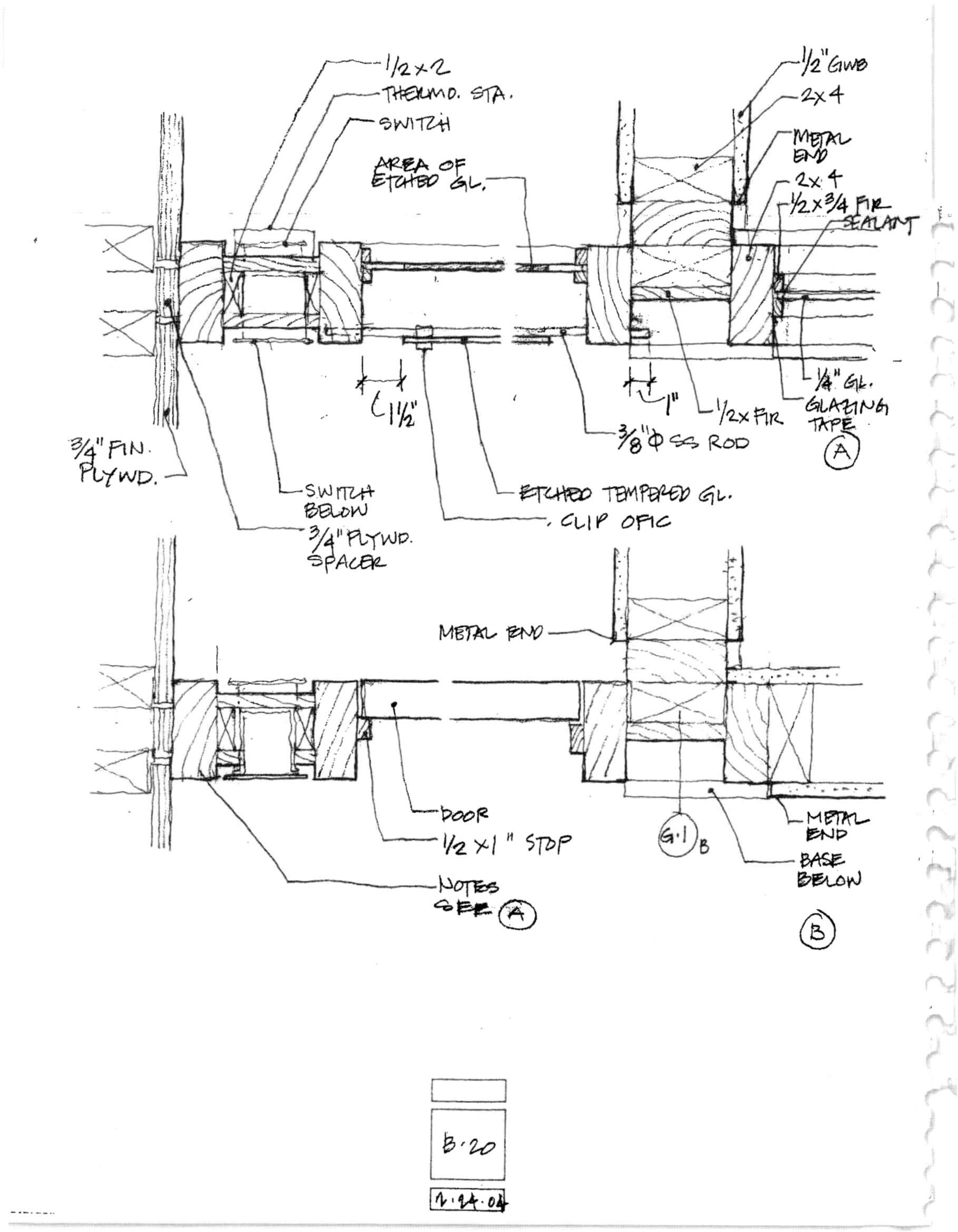

SPECIFICATION BOOK DRAWINGS BY CUTLER ANDERSON ARCHITECTS

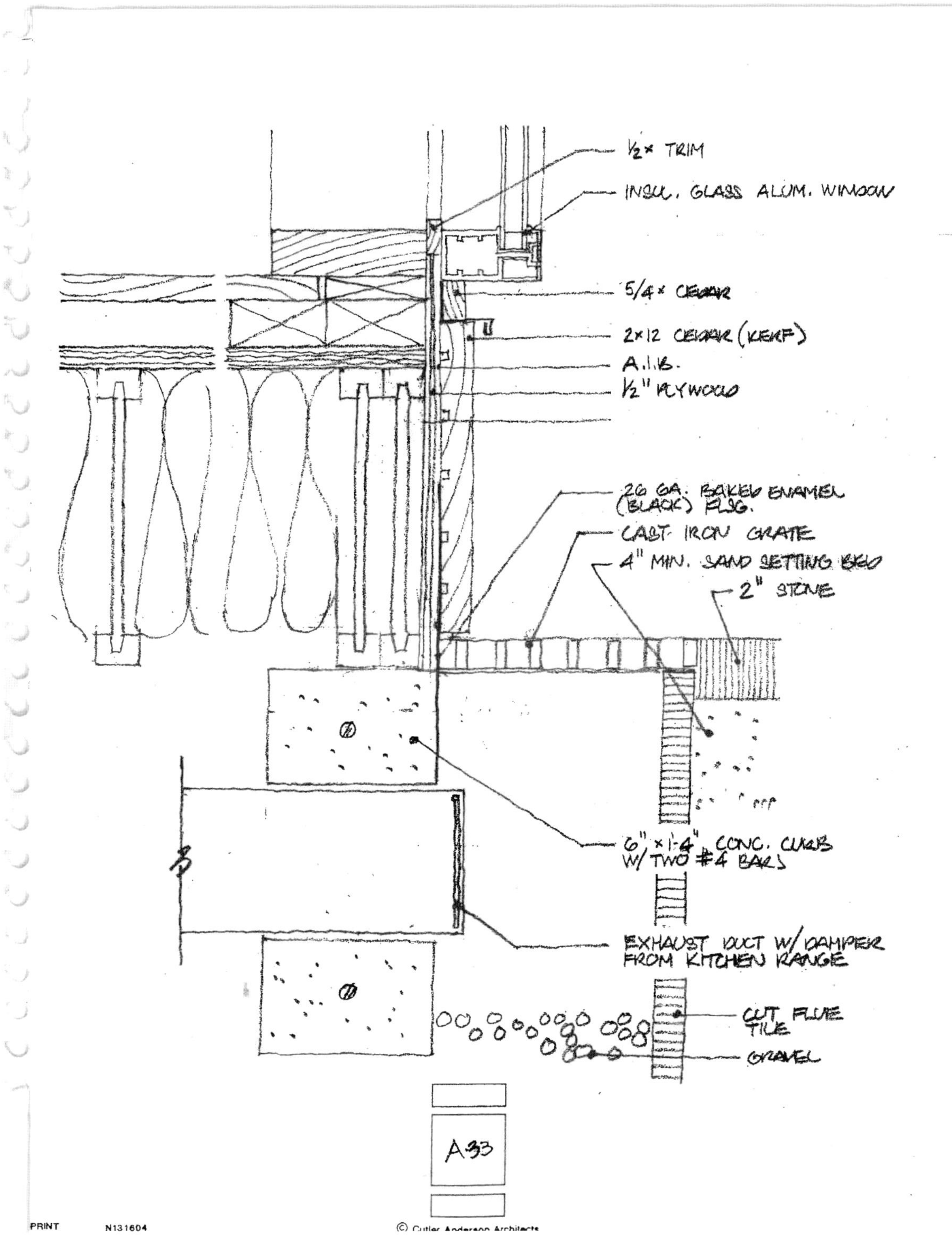
½× TRIM
INSUL. GLASS ALUM. WINDOW
5/4× CEDAR
2×12 CEDAR (KERF)
A.I.B.
½" PLYWOOD
26 GA. BAKED ENAMEL (BLACK) FLG.
CAST IRON GRATE
4" MIN. SAND SETTING BED
2" STONE
6"×1'-4" CONC. CURB W/ TWO #4 BARS
EXHAUST DUCT W/ DAMPER FROM KITCHEN RANGE
CUT FLUE TILE
GRAVEL
A33
PRINT
N131604
© Cutler Anderson Architects

do without the "good stuff" like copper, brass, mahogany, slate, or engineered lumber. Low budgets require smart planning, clever material choices, and efficient labor on the builder's side.

My experiences as a builder and client (I've done several personal projects) have taught me many building lessons along the way.

Mock-ups are critical to give the owners a real look at their facades and interiors. Sometimes they are expensive to put together, but they can prevent problems from developing later.

Try to include outside basement access, such as a Bilco door. If there are mechanicals down there, then tradesmen need to reach them without going through the front door. Likewise, if there is a gabled attic, try to squeeze in a stairway somewhere instead of a pull-down unit. They are more user-friendly, making the space more useful for access and storage.

Location siting layout is obviously critical for drainage, landscaping, and daylight exposure. Getting the elevations right is essential: almost always, higher is better. I am in favor of roof overhangs, though I know many designers prefer little or none. I think of it as a bill on a cap—it protects the face and skin from exposure to the elements. On a house, it should also keep water away from the structure.

The minimum size for a bedroom door should be 32 inches, but 34 to 36 inches is preferable, to accommodate furniture. It is also helpful to make hallway widths a few inches wider. The ideal stair-riser dimension is 7 inches or less, with a tread depth of 11 or 12 inches. Staircases can be

ARCHITECT AND BUILDER CONSULTING TOGETHER IS THE BEST WAY TO FIND SOLUTIONS.

hard to fit into a plan, but I dislike steep rises. For exterior staircases, I aim for 6-inch risers and a deep 18-inch tread depth, especially when building in stone.

Bathroom layout can make a real difference, especially to us older people, who like grab bars near and inside tubs and showers, and anti-slip traction surfaces on floors. I once installed a semi-round decorative bullnose tile around a tub. The mitred outside corner was lethally sharp and dangerous. If there are kids to consider, ease or blunt corners wherever possible, even on furniture. Timer switches are a good idea for bathroom exhaust fans, or at least separate switches for lights and fans, as these can be left on for hours and vacuum out indoor air.

Fireplace hearths should be raised in bedrooms so they can be seen from the bed, and on upper floors they should be gas—who's carrying wood upstairs? I had a client complain about a supply air duct blowing down directly over his bed. Reviewing locations helps find optimum placement. Usually, HVAC people want to curtain the room-exterior perimeters with supply air and return ducts to the interiors. I've seen some older homes with up-and-down return air ducts on a wall—open the upper for summer, and the lower for winter.

Mechanical rooms need space! We built a very contemporary building, complex mechanically and otherwise, and the equipment was squeezed into a space that felt like being in a submarine. It is useful to have an empty conduit run from the mechanical area to the attic, facilitating access for future needs. Also, extra conduits are used when running electrical and water lines as backups.

Architecturally ambitious details are wonderful, but can sometimes ignore material behavior: shrinkage, swelling, and movement. Architect and builder consulting together is the best way to find solutions.

MIRROR HOUSE

In nearby Pound Ridge, NY, we completed an intricate building with KieranTimberlake architects. The site is a challenging outcropping of rock ledge, and the foundation is nestled on and between massive boulders. The exterior skin comprises mirrored stainless-steel and zinc panels, and the highly reflective surfaces appear almost futuristic. Wonderful, thoughtful clients, a great design team, and some impressive performances by our employees made this a superlative project with a happy result.

ARCHITECTURE: KIERANTIMBERLAKE

WESTCHESTER RETREAT

This is one of my personal favorites, which I often describe as a contemporary mansion. Building this elegant home, with sumptuous spaces inside and out, was an absolute pleasure, working with Jim Olson and William Franklin of Olson Kundig. Not to be overlooked, the owners should be honored with a "nicest people to work for" award. The planning and plans evidently took years to complete before we ever got involved and were very thorough: 300 pages . . .

ARCHITECTURE: OLSON KUNDIG

MODERN FARMHOUSE

In the Hudson Valley, I was privileged to work with an impressively talented, India-based architect named Niels Schoenfelder of Mancini Design. A German man who had worked in Paris, he relocated to India where he designed a house for a South African woman who later purchased a large bucolic farm property with multiple buildings. The plan involved a new farmhouse-style contemporary building with elaborate structural details, interior finishes, and room layouts. The owner and architect knew exactly the look they wanted, and it required four or five stone veneer mock-ups to get it right. The unique ambience of this home radiates joyfulness. The style and sensibility are like no other I've been involved in.

ARCHITECTURE: MANCINI DESIGN

FOUR POINTS HOUSE

This new build on a ridge offers beautiful views in bucolic Sharon, CT. The unique geometric shape is clad in burned wood siding to achieve the dark exterior color. Architect Martin Finio and his team, the interiors firm, and the landscape architects were great to work with, creating a well-conceived plan placed in a gorgeous setting.

"The primary driver of this house's design was the four distinct views it frames: a long mountain view to the west, a meadow view to the south, a farm view to the north, and a forest view to the east. With its U-shaped plan, the house forms three walls of a courtyard with the treeline of the forest forming the fourth. The exterior is clad in cedar processed in the Shou Sugi Ban method, which chars the wood to protect against fire and the elements and results in a rich black color and scaly texture. Inside, the exposed framing is made of mass timber. The primary suite occupies the south volume of the house, with a long wing of social, entertaining, and dining spaces between it and the guest suite to the north."

—MARTIN FINIO

ARCHITECTURE: IN STUDIO ARCHITECTURE

FOR THE BUILDERS

When it comes to hiring, I have made a lot of mistakes and, unfortunately, could list my "Hall of Shame" candidates—which I won't. I also made many excellent choices, resulting in some great teams, particularly now: we have around 35 employees and not a weak link among them. I hire based on instincts and references. Having experience in what we do is great, but not imperative. Smart, motivated people can learn fast and sometimes be easier to train.

References are important to confirm your instincts and weed out impostors. I'm always leery if a résumé indicates a pattern of two years here and there; I want to know why someone can't settle into a job. I preach and practice that loyalty goes both ways. If things slow down or people need time off, we have their back.

I got smart a few years back and designated a star employee named Jack Truman as Company President, and eventually he became a part owner. Hiring architects and architect graduates has allowed us to better communicate with designers and, coupled with experienced, savvy carpenter builders, provides our clients with a complete service. We staff large projects with an office-based project manager, a dedicated job-site supervisor, and usually a laborer or handy person. Responsibilities are separated, but we all work together, including assistant project managers and senior management.

Collecting, paying, charging … money is a big subject.

I was once owed over $3,000,000, and when the owner's representative asked: "What are you worried about, Dave?" I told him: "We are not IBM. This is a small business." Builders are almost always in debt, as we usually do a month's work, then write a bill, and then commonly wait another month to get paid. There are exceptions, of course, but generally the fastest payers get the fastest service. We do seek a deposit near the job's outset, but I've never been comfortable asking for a high percentage, as I don't want to make the client uneasy either. So it's a delicate balance. We do some research on potential clients, but as Billy Joel sang: "It's a Matter of Trust." It's got to be there and earned. There's almost always some drama regarding final payments. I had one good client ask me almost tearfully, if we really would come back to honor our warranty. But try not to get into arguments with them as they are the boss and usually don't want to hear they're wrong, even when they are.

So many of our projects never end, it seems. Late-arriving pieces, touch-ups, additions, and punch-list work can go on for months, draining everyone. Sometimes we swap out our fatigued job supervisor at 90 percent completion to get some fresh eyes and energy in charge. One Georgian Colonial brick house underwent significant remodeling almost immediately after we completed it. After living in it for a month, the owner decided that many design "flaws" had to be corrected, so we were back at it, ripping and tearing things apart soon after.

If you want loyalty and responsiveness from your subcontractors and suppliers, then be fair and pay them promptly. My wife and I never bought anything unless we absolutely needed or required it, so we always remained solvent. We were not the

IF YOU WANT LOYALTY AND RESPONSIVENESS FROM YOUR SUBCONTRACTORS AND SUPPLIERS, THEN BE FAIR AND PAY THEM PROMPTLY.

company driving a fleet of new trucks and we still operate in much the same frugal way. I would suggest extreme caution about outlaying large deposits, which independent window companies and many millwork shops request. Why do they need 50 percent to commence an order? We did have a West Coast window company go bankrupt with our $40,000 deposit, and we were powerless to recover it. Also, a roofing contractor I used on occasion ran away with a materials advance, then disappeared. I'd make him disappear again if I could find him. Like any business, it's about being smart, fair, and honorable with coworkers and clients, establishing credit and credibility throughout your industry network.

I TELL OUR PEOPLE NOBODY WAKES UP IN THE MORNING AND DECIDES TO HAVE A TABLE-SAW ACCIDENT OR FALL OFF A LADDER. IT'S ALWAYS UNPLANNED, UNFORESEEN, AND SPONTANEOUS.

I've got lots of stories, but too many to mention here. I'll save that for another book, but I'll discuss rules 1 and 2: protection of people and property. I tell our people nobody wakes up in the morning and decides to have a table-saw accident or fall off a ladder. It's always unplanned, unforeseen, and spontaneous.

The owner has entrusted the builder to protect and preserve the property. That involves monitoring who is working or visiting there, and making sure a high level of job-site safety rules are in place, respected, and enforced. We have a qualified outside safety consultant make monthly visits to all jobs and write detailed follow-up reports to be reviewed by all concerned. I am petrified of a job-site fire, and we have had some close calls. **A building is never more vulnerable to destructive fire than when under construction.** The biggest dangers are smoking, halogen lightbulbs, sawdust piles, propane heaters, open torch work, combustibles (paint thinners, oils, propane, etc.) stored in the building—just demand none! Floor-sanding machines must be emptied of dust daily and ideally stored outside or in a garage overnight. Frayed power cords are among the many things to be mindful of. It is always a good idea to activate alarm systems, smoke/heat detectors as soon as possible, and linked to a central station if possible. I like to have a charged water hose in the middle of the house and, of course, conventional portable extinguishers lying around, and people who know how to use them. **You cannot be too careful here.** One of my first ever jobs was working under a contractor who had me removing paint from clapboards with a handheld propane torch. Sensing danger, I wisely filled a bucket of water and placed it next to my ladder. A piece of felt paper ignited, and I was able to quickly douse it; otherwise, the Brewster, MA, original town hall wouldn't be standing any longer. I didn't work for that guy for very long.

I've heard dozens of nightmare stories—a house painter turned the exterior, corner-mounted twin flood lights to face upward, away from his work; when the lights were turned on by a timer switch, they ignited an overflowing pile of pine needles in the adjacent gutter—and the house burned down. In another case, the owner of a brand-new Greenwich McMansion requested the floor sander to wrap the large new crystal chandelier with plastic to protect it from dust. The light was mistakenly turned on and, soon after, melting plastic droplets ignited the wet floor finish—house gone. Be vigilant.

Maintenance work has become a large and growing part of our business. These houses are complex, and new homeowners need guidance in maintaining and using them. We have great clients, and servicing them can be lucrative and fill gaps in our schedule. We've gotten very good at providing owner's manuals at job completion. It's a lot of work collecting catalogues and purchase orders, and collating information into a useful, readable format. And have your mechanical subcontractors label everything legibly, including dog tags on plumbing shutoffs and electrical panels. The owner absolutely deserves for this to be done right.

Self-promotion is important for builders. If you've got some good work to show off, do so. Good photography is essential. You build a business reputation on client references—those must be earned, and a contented customer will gladly provide one.

HILLTOP HOUSE

It appears as if a serene, quiet, low-profile building has been dropped onto a windy grassy hilltop. We were privileged to work with two great design architects from Bohlin Cywinski Jackson, and a terrific and daring New York client who had acquired this multi-acre property years earlier. Exposed on all sides to the elements, this stylistically reserved house contains all the requisite creature comforts of a modern contemporary home. The living experience is feeling immersed in nature with spectacular views, yet hardly disturbing the surrounding countryside.

ARCHITECTURE: BOHLIN CYWINSKI JACKSON

STONE CONNECTICUT MANOR HOUSE

It is indeed the stonework that stands out on this manor house. We had a great team chinking and fitting fieldstone for many months. I wanted to keep the miniature mock-up. Steven Harris Architects, and the interiors team, Pierre Yovanovitch from France, created some fantastic room designs and finishes on a beautiful, expansive site.

ARCHITECTURE: STEVEN HARRIS ARCHITECTS

TEAHOUSE

The Teahouse is a one-bedroom house built for a prominent fashion designer. It's beautiful and has always impressed me as a complete thought, as the spaces seem so intricately and seamlessly connected. The decor and furnishings are eclectic and artful. The exterior is quiet yet graceful and compelling. The architectural team of Calvin Tsao and Zach McKown has created something extraordinary here.

ARCHITECTURE: TSAO & MCKOWN ARCHITECTS

NEW CANAAN MODERN

This was my third personal property project. The existing structure was torn down and rebuilt in the exact same footprint, recycling only a small part of the foundation. The new building mimics the original but is wholly re-envisioned and modernized from the 1950s conception. The new second floor seems to float above the first, paying homage to the original Eliot Noyes creation.

The spectacular finished new home on this gorgeous property was widely published, and changed the direction of new business for our company.

ARCHITECTURE: JOEB MOORE & PARTNERS

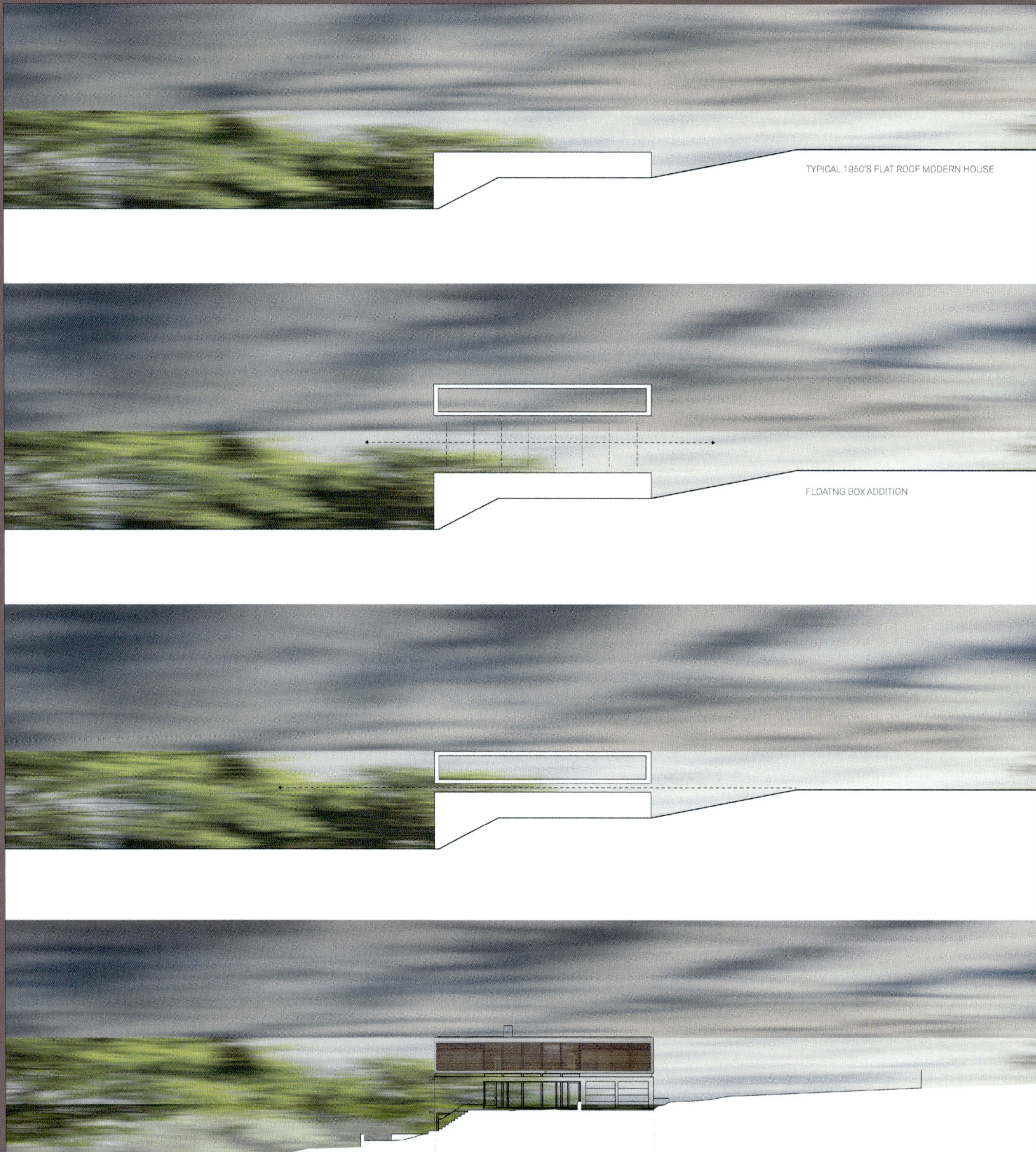

FLOATING BOX ADDITION ELEVATION BY JOEB MOORE & PARTNERS

NORTH VIEW \ ENTRY APPROACH OF ORIGINAL ELIOT NOYES HOUSE

SOUTHEAST VIEW \ REAR YARD OF ORIGINAL ELIOT NOYES HOUSE

MORE HOUSES

GREENWICH WATERFRONT

Joeb Moore & Partners revisualized a standard McMansion in a prime location and converted it into a comfortable home with a handsome exterior. Major renovation took place throughout the interior spaces, and extensive landscaping was done around the large new pool. Once again, we were lucky to have a great client.

ARCHITECTURE: JOEB MOORE & PARTNERS

LOST & FOUND LAB

COS COB \ CONNECTICUT

ARCHITECTURE: JOEB MOORE & PARTNERS

GREENWICH 2008

This big project was on a large lot near the village center. Water flowed down onto and across this property, requiring major well-engineered diversion work. Water being a theme here, the primary entry involved stepping onto stone tablets above a reflection pool that spilled over a 10-foot waterfall before the water was recycled into a container visible from the stairway below. Building such a design was definitely a first for us. This was a large house for a large family, and they got what they wanted, including a kids' homework room and a maid's quarters. The three-story glass stairwell is impressive and overlooks an exterior deck made from vertically placed two-by-fours, so it will likely be solid for decades.

ARCHITECTURE: JOEB MOORE & PARTNERS

MARCEL BREUER HOUSE REMODEL

This was an expansion/renovation of a prominent architect and furniture designer's New Canaan, CT, residence. Architect Toshiko Mori and her team designed a stunning glass entry and a new bedroom wing cantilevered over the rear slope of the property. Stonework was painstakingly disassembled and reinstalled, including the entire mosaic flagstone interior floor, to replicate the original after removal for mechanical placement. Work included an overhaul and expansion of a pool house and patios.

ARCHITECTURE: TOSHIKO MORI ARCHITECT

NEW CANAAN LEISURE

This wonderful project gave us the opportunity to work with a fun, adventurous couple, the talented architect Scott Specht from Texas, and a great interiors team. Spacious rooms were specifically designed for the couple's lifestyle, including an amazing pool house and a concrete outdoor lounge featuring a firepit.

ARCHITECTURE: SPECHT NOVAK ARCHITECTS

STONINGTON WATERFRONT

STONINGTON \ CONNECTICUT

ARCHITECTURE: JOEB MOORE & PARTNERS

CP HOUSE \ AUSTERLITZ, NY

LONG ISLAND WATERFRONT \ WESTPORT, CT

GREENWICH MODERN \ GREENWICH, CT

CONNECTICUT MODERN RENOVATION \ WILTON, CT

GREENWICH MODERN \ GREENWICH, CT

WESTCHESTER ADDITION \ WESTCHESTER, NY

RYE WATERSIDE \ RYE, NY

WESTCHESTER RESIDENCE \ WESTCHESTER, NY

MILLERTON MOUNTAINTOP \ MILLERTON, NY

AFTERWORD

There are some fabulous buildings in this collection and more underway. It's personally very satisfying to have built a team that gets it done, and to have met and worked with so many talented professionals—architects, designers, tradespeople, engineers, and more. And, of course, the clients provide the impetus and the means to make it happen. It starts with them.

I have been seeking an expressive, creative outlet most of my life, and I'm lucky to have found it. Several years ago, I told the team that while we're not Microsoft or Apple, we have changed the world a bit in our own modest way by bringing these great designs to life. Having visited Europe several times recently, I am constantly stunned and humbled looking at amazing, complex, and decorative buildings built to perfection hundreds of years ago, without the modern tools and equipment we use today.

With building sciences and modern technology evolving rapidly, we can expect dramatic changes ahead in construction: alternative energy sources, prefabricated homes, and 3D-printed buildings. It would certainly be nice to see developments in providing mass-produced, low-cost housing for underdeveloped countries and for homeless people everywhere, including in the USA. America has typically been a world leader in spreading goodwill using the "soft power" of sharing technological advances, by donating food and healthcare, and in the work of private charities such as Habitat for Humanity, for which Jimmy Carter volunteered for more than 35 years.

Thanks for reading. Peace, David L. Prutting

AWARDS

2025 AIA PHILADELPHIA MERIT AWARD
Project: Hilltop House
Architect: Bohlin Cywinski Jackson

2025 HOBI AWARD, HOME BUILDERS ASSOCIATION OF CONNECTICUT, BEST MODERN CUSTOM HOME
Project: Greenwich Harbor House
Architect: Amanda Martocchio

2024 MOFFLY MEDIA A-LIST AWARD, BEST MODERN ARCHITECTURE
Project: Four Points House
Architect: In Studio Architecture

2023 AMERICAN ARCHITECTURE AWARD
Project: Four Points House
Architect: In Studio Architecture

2023 MOFFLY MEDIA, ATHOME A-LIST AWARD, BEST ENTRYWAY
Project: Indian Head Residence
Architect: DB Design

2022 HOBI AWARD, HOME BUILDERS ASSOCIATION OF CONNECTICUT, BEST CUSTOM HOME COMBINING OLD & NEW
Project: Modern Farmhouse
Architect: Niels Schoenfelder for Mancini Design

2022 AIA CT DESIGN EXCELLENCE AWARD
Project: Lost & Found Lab
Architect: Joeb Moore & Partners

2021 HOBI AWARD, HOME BUILDERS ASSOCIATION OF FAIRFIELD COUNTY, BEST MODERN CUSTOM HOME OVER 5000SF
Project: Greenwich Modern
Architect: Joeb Moore & Partners

2019 MOFFLY MEDIA A-LIST AWARD, BEST RENOVATION
Project: Modern Barn
Architect: Specht Novak Architects

2018 MOFFLY MEDIA A-LIST AWARD, BEST MODERN RENOVATION
Project: New Canaan Residence
Architect: Specht Novak Architects

2018 AIA CT DESIGN AWARD, EXCELLENCE IN PRESERVATION, RESTORATION, OR ADAPTATION

2018 HOBI AWARD, HOME BUILDERS ASSOCIATION OF CONNECTICUT, BEST RESIDENTIAL REMODEL

2018 HOBI AWARD, HOME BUILDERS ASSOCIATION OF CONNECTICUT, REMODELED HOME OF THE YEAR
Project: River House
Architect: Joeb Moore & Partners

2018 CUSTOM HOME BUILDER DIGEST, BEST LUXURY HOME BUILDER IN CONNECTICUT

2017 INTERIOR DESIGN MAGAZINE BEST OF YEAR AWARD, COUNTRY HOUSE
Project: Teahouse
Architect: Tsao & McKown Architects

2017 BUILDER'S CHOICE & CUSTOM HOME DESIGN AWARD, PROJECT OF THE YEAR
Project: Mirror House
Architect: KieranTimberlake

2017 MOFFLY MEDIA A-LIST AWARD, BEST MODERN ARCHITECTURE
Project: Mirror House
Architect: KieranTimberlake

2016 AIA NY DESIGN AWARD, LARGE RESIDENTIAL
Project: Pound Ridge House
Architect: KieranTimberlake

2016 HOBI AWARD, HOME BUILDERS & REMODELERS ASSOCIATION OF FAIRFIELD COUNTY, BEST MODERN STYLE CUSTOM HOME
Project: Long Island Waterfront
Architect: Sellars Lathrop Architects

2016 AIA NEW ENGLAND CITATION AWARD, HISTORIC PRESERVATION & ADAPTIVE REUSE
Project: Stonington Waterfront
Architect: Joeb Moore & Partners

2016 CT COTTAGES & GARDENS INNOVATION IN DESIGN AWARD, BUILDER RECOGNITION
Project: Long Island Waterfront
Architect: Sellars Lathrop Architects

2016 LUXE MAGAZINE GOLD LIST HONOREE
Project: New Canaan Transitional
Architect: Amanda Martocchio Architecture

2015 AIA CT DESIGN AWARD

2015 AIA WESTCHESTER & HUDSON VALLEY DESIGN AWARD
Project: Westchester Addition
Architect: Joeb Moore & Partners

2015 AIA WESTCHESTER & HUDSON VALLEY DESIGN AWARD
Project: Hilltop Modern
Architect: Deborah Berke & Partners

2014 CT COTTAGES & GARDENS INNOVATION IN DESIGN AWARD, BUILDER RECOGNITION; ARCHITECTURE

2014 MOFFLY MEDIA A-LIST AWARD, BEST MODERN ARCHITECTURE

2014 DESIGN AWARD, BOSTON SOCIETY OF ARCHITECTS

2013 AIA CT DESIGN AWARD, RESIDENTIAL MERIT AWARD
Project: Stonington Waterfront
Architect: Joeb Moore & Partners

2013 HOBI AWARD, HOME BUILDERS ASSOCIATION OF CONNECTICUT, OUTSTANDING CUSTOM HOME, 5000–6000SF
Project: Stonington Waterfront
Architect: Joeb Moore & Partners

2013 HOBI AWARD, HOME BUILDERS ASSOCIATION OF CONNECTICUT, BEST CONTEMPORARY HOME REMODEL
Project: Greenwich Waterfront
Architect: Joeb Moore & Partners

2014 AIA NY MERIT AWARD
Project: Ghent Residence
Architect: Toshiko Mori Architect

2014 HOME BUILDERS & REMODELERS ASSOCIATION OF FAIRFIELD COUNTY, REMODELER OF THE YEAR

2013 AIA AUSTIN HONOR AWARD

2013 MOFFLY A-LIST AWARD, BEST MODERN ARCHITECTURE

2013 MOFFLY MEDIA A-LIST AWARD, BEST POOL HOUSE
Project: New Canaan Residence
Architect: Specht Novak Architects

2011 AIA PORTLAND'S ARCHITECTURE + DESIGN FESTIVAL, DESIGN MATTERS 2001: A TOUR OF EXCEPTIONAL PORTLAND HOMES
Project: Leedy House
Architect: Paul McKean Architecture

2011 AIA CT DESIGN AWARD, SINGLE-FAMILY RESIDENTIAL PROJECT
Project: New Canaan Townhouse
Architect: Joeb Moore & Partners

2011 HOBI AWARD, HOME BUILDERS ASSOCIATION OF CONNECTICUT, BEST IN-TOWN CUSTOM HOME / BEST SPEC HOME
Project: New Canaan Townhouse
Architect: Joeb Moore & Partners

2010 CUSTOM HOME DESIGN AWARDS, CUSTOM HOME MAGAZINE, GRAND AWARD

2010 TUCKER DESIGN AWARD, BUILDING STONE INSTITUTE, EXCELLENCE IN NATURAL STONE DESIGN
Project: New Canaan Pool House
Architect: Amanda Martocchio Architecture

2009 AIA NEW ENGLAND MERIT AWARD
Project: Greenwich 2008
Architect: Joeb Moore & Partners

2008 CUSTOM HOME DESIGN AWARDS, CUSTOM HOME MAGAZINE, MERIT AWARD
Project: Crafted Sanctuary
Architect: Cutler Anderson Architects

2001 AIA PORTLAND'S ARCHITECTURE + DESIGN FESTIVAL, DESIGN MATTERS 2001: A TOUR OF EXCEPTIONAL PORTLAND HOMES
Project: Cape Cod House
Architect: Benjamin Waechter

FEATURED HOUSES

HORIZON HOUSE PP32–39
Architecture: Steven Holl Architects
Landscape: Hollander Design
Photography: Paul Warchol

GREENWICH HARBOR HOUSE PP40–49
Architecture: Amanda Martocchio Architecture
Landscape: James Doyle Design
Photography: Michael Biondo

A HOUSE IN GHENT PP50–61
Architecture: Toshiko Mori Architect
Landscape: Edwina von Gal
Photography: Thomas Loof pp50–51, 52–53, 54, 56, 59, 60–61; Paul Warchol p57

RIVER HOUSE PP62–73
Architecture: Joeb Moore & Partners
Landscape: Devore Associates
Photography: Timothy Schenck

NEW CANAAN TOWNHOUSE PP74–85
Architecture: Joeb Moore & Partners
Landscape: Richard Bergman
Photography: David Sundberg / Esto Photography pp74–75, 76, 77, 78, 79, 81, 82–83, 84, 85; Michael Biondo p82

MIRROR HOUSE PP100–109
Architecture: KieranTimberlake
Landscape: Reed Hilderbrand
Photography: Peter Aaron / Esto Photography

WESTCHESTER RETREAT PP110–17
Architecture: Olson Kundig
Landscape: Hollander Design
Photography: Aaron Leitz pp110–11, 114–15; Nic Lehoux pp112, 116–17; Michael Biondo p113

MODERN FARMHOUSE PP118–31
Architecture: Mancini Design
Landscape: Mancini Design
Interiors: Mancini Design
Photography: Bjorn Wallander

FOUR POINTS HOUSE PP132–41
Architecture: In Studio Architecture
Landscape: Kathryn Herman Design
Interiors: Rees Roberts + Partners
Photography: David Mitchell

HILLTOP HOUSE PP152–61
Architecture: Bohlin Cywinski Jackson
Photography: Nic Lehoux

STONE CONNECTICUT MANOR HOUSE PP162–71
Architecture: Steven Harris Architects
Landscape: Bold Meadows
Interiors: Pierre Yovanovitch
Photography: Stephen Kent Johnson

TEAHOUSE PP172–83
Architecture: Tsao & McKown Architects
Landscape: Tsao & McKown Architects
Interiors: Tsao & McKown Architects
Photography: Michael Biondo pp173, 178–79, 180–81, 182–83; Simon Upton pp174–75, 176, 177

NEW CANAAN MODERN PP84–93
Architecture: Joeb Moore & Partners
Landscape: Reed Hildebrand
Photography: David Sundberg / Esto Photography pp184–85, 186, 187, 190–91, 192–3, 193; Courtesy of Prutting + Co. p189

MORE HOUSES

GREENWICH WATERFRONT PP196–99
Architect: Joeb Moore & Partners
Landscape: Hollander Design
Photography: David Sundberg / Esto Photography

LOST & FOUND LAB PP200–203
Architect: Joeb Moore & Partners
Photography: Michael Biondo

GREENWICH 2008 PP204–207
Architect: Joeb Moore & Partners
Landscape: Reed Diller & Associates
Photography: David Sundberg / Esto Photography

MARCEL BREUER HOUSE REMODEL PP208–11
Architect: Toshiko Mori Architect
Landscape: Quennell Rothschild & Partners
Photography: Michael Biondo pp208–209, 210, 211 (right); Paul Warchol p211 (left)

NEW CANAAN LEISURE PP212–15
Architect: Specht Novak Architects
Landscape: Gunn Landscapes
Interiors: Carrier & Co.
Photography: Elizabeth Felicella Photography

STONINGTON WATERFRONT PP216–19
Architect: Joeb Moore & Partners
Landscape: Reed Hildebrand
Photography: David Sundberg / Esto Photography

PHOTOGRAPHY

Michael Biondo: pp24, 26 (Toshiko Mori Architect; Marcel Breuer Remodel); pp96–97 (Gertler & Wente Architects; Lakeside Berkshires); p220 (Toshiko Mori Architect; CP House); pp222–23 (Bassam Fellows Architects; New Canaan Residence); p226 (Doug Patt; Rye Waterside); pp230–31 (280SR)

Esto Photographics: p10 (Joeb Moore & Partners; Westchester Renovation)

Grice Photography: pp88, 90–91 (Cutler Anderson Architects; Weekend Retreat); p144 (Cutler Anderson Architects; Crafted Sanctuary)

Amanda Kirkpatrick: p23 (Rafe Churchill; Larchmont Traditional)

Nic Lehoux: pp2, 30 (Olson Kundig; Westchester Retreat); pp246–47 (Bohlin Cywinski Jackson; Hilltop House)

Ines Leong, Archphoto: p244 (Zehra Kuz, Oasis Design Lab; Wilton Modern)

Michael Moran: p12 (Specht Novak Architects; Modern Barn)

Ren Nickson Photography: p242 (Cutler-Anderson Architects; McNeil Residence)

Eric Petscheck: p224 (Emily Lindberg Design; Connecticut Modern Renovation)

Courtesy of Prutting + Co.: pp18, 19, 189

Courtesy of Prutting Family: pp6, 8, 9

David X. Prutting: p15 (inlaid floor detail); p16 (Joeb Moore & Partners; Greenwich Leisure); p17 (wood detail); p21 (Robertson Landers Architects; Craftsman Style Home); pp22, 238 (Mark Demmerle Architects; Shingle Style Home); p227 (bottom; Steven Harris Architects; Millerton Mountaintop)

Jason Schmidt: pp147, 240–41 (Deborah Berke + Partners; Hilltop Modern)

David Sundberg / Esto Photography: p4, (Joeb Moore & Partners; Greenwich 2008); p29 (Joeb Moore & Partners; New Canaan Townhouse); pp86, 151, 221 (bottom), 225 (left; Joeb Moore & Partners; Greenwich Modern); pp194, p225 (right; Joeb Moore & Partners; Westchester Addition)

Nicholas Venezia: pp95, 227 (top; Selldorf Architects; Westchester Residence)

Bjorn Wallander: pp98, 232 (Mancini Design; Modern Farmhouse)

Paul Warchol: pp142, 148 (Toshiko Mori Architect; Marcel Breuer Remodel)

Matthew Williams: p221 (top; Sellers Lathrop Architects; Long Island Waterfront)

Elizabeth Felicella Photography: p236 (Specht Novak Architects; New Canaan Leisure)

ART DIRECTION AND BOOK DESIGN
Yvonne Teresa Creative Studio

CURRENT STAFF AND EMPLOYEES

The people who GET IT DONE.

PRESIDENT & PARTNER: Jack Truman

CHIEF FINANCIAL OFFICER: Frank DiMicco

SENIOR PROJECT MANAGERS: Rick Bellefueille, Heath Horn

DIRECTOR OF SERVICE: Stephanie Rende

OFFICE MANAGER: Daniel Callahan

Dave Berndt

Nic Berry

Jon Brindisi

Louis Brugnatelli

Mark Burtnett

Lauren Cali

Evan Cronly

Carlos Cubero

Sean DiMicco

Sean Flood

Trey Gregory

Kathryn Grove

Peter Hoyt

Eric Jones

Taylor Leary

Craig Lynch

Dave McGeehan

Chris McIntyre

Deric Mizokami

Jim Newman

Tim Nullet

Kevin Poole

Vernard Ramirez

Marek Sernicki

Nicole Sikoryak

Andy Stryker

Jonathan Sweibel

Tom Tryczynski

Peter Tvrdik

PAST CONTRIBUTORS & ALLIES

Special THANK YOU to the many talented people who have contributed greatly to our success.

Chyngyz Asanov
Anton Balek
Frank Bedini
Dave Benore
Gerber Bonilla
Bruce & Don Bottecelli
Luppino Brothers
Caesar Cali
Pete Callahan
Frank Candullo
Jose Carranza
Jose Castillo

Ian Dowell
Joe Ferro
John Finlay
Andrew Friedman
Ed Geoghegan
Ralf Klemets
Rob Mackey
Anthony Manca
Al Mansi
Daniel Mendelson
Doug Miller
Bill Nietsch

Michael O'Connor
Al Petrucci
Jon Pompea
Al Pote
Leonard Richall
Daryl Sande
Jeff Scalise
Piotr Sernicki
John St. John
Simon Stigsby
Curt Stratton
Ed Turlis

Published in Australia in 2026 by
The Images Publishing Group Pty Ltd
ABN 89 059 734 431

Offices
MELBOURNE
Waterman Business Centre
Suite 64, Level 2 UL40
1341 Dandenong Road
Chadstone, Victoria 3148
Australia
Tel: +61 3 8564 8122

NEW YORK
6 West 18th Street 4B
New York, NY 10011
United States
Tel: +1 212 645 1111

SHANGHAI
6F, Building C, 838 Guangji Road
Hongkou District, Shanghai 200434
China
Tel: +86 021 31260822

books@imagespublishing.com
www.imagespublishing.com

The Images Publishing Group Reference Number: IM1778

NATIONAL LIBRARY OF AUSTRALIA

A catalogue record for this book is available from the National Library of Australia

Title: A Builder's Life Done Well: The Story of Prutting + Co.
Author: David L. Prutting
ISBN: 9781923094277

This title was commissioned in IMAGES' Melbourne office and produced as follows:
Art Direction and Book Design: Yvonne Teresa Creative Studio; *Assets Coordination* Dan Callahan; *Editorial* Rebecca Gross, Helen Koehne, Jeanette Wall; *Final Art* Nicole Boehringer; *Production* Simon Walsh, Heather Johnson; *Proofing* Kate James; *Prepress* Splitting Image (Melbourne)

EU GPSR Authorised Representative: Easy Access System Europe Oü
Company Registration ID: 16879218 | Address: Mustamäe tee 50, 10621 Tallinn, Estonia
Email: gpsr@easproject.com | Tel: +358 40 500 3575

FSC www.fsc.org MIX Paper | Supporting responsible forestry FSC® C015829

Printed on 150gsm Magno Matt art paper (FSC) in Italy by Printer Trento

PRUTTING + CO.

75 CAMP AVENUE
STAMFORD, CT 06907

PRUTTING.COM